SCIENCE WORKS

STAYING ALIVE

THE STORY OF A FOOD CHAIN

IT'S SIMPLY MOUTH-WATERING!

Jacqui Bailey Matthew Lilly

A & C BLACK • LONDON

It was the middle of the day in the African savannah, and the sun's heat baked the land like an oven.

Sunlight sparkled on the surface of a small lake. Every now and then, the snout of a hippopotamus bobbed up and down in the water.

Savannah is the name given to the hot, tropical grasslands found in parts of Africa, Asia, Australia and South America. Different types of savannah have different animals and plants living in them.

A short distance from the lake, a tree stood among the grasses and shrubs. The sun blazed down on the tree.

The tree needed the sunlight. Without it, the tree could not grow and make its food — and without food the tree would die.

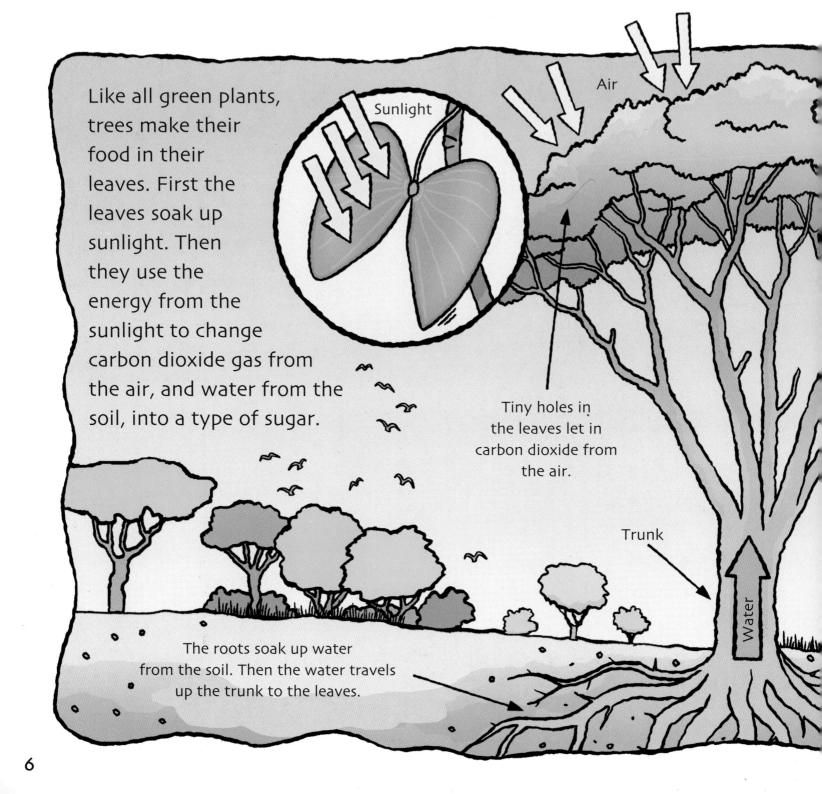

Like all green plants, trees make their food in their leaves. First the leaves soak up sunlight. Then they use the energy from the sunlight to change carbon dioxide gas from the air, and water from the soil, into a type of sugar.

Sunlight

Air

Tiny holes in the leaves let in carbon dioxide from the air.

Trunk

Water

The roots soak up water from the soil. Then the water travels up the trunk to the leaves.

I'M A FOOD FACTORY.

SLURP!

SLURP!

Well, not quite. To be strong and healthy, trees also need tiny bits of chemicals called minerals. The minerals are in the soil and the trees suck them up, with the water, through their roots.

Trees use the sugary food they make to give them the energy to grow and to make more leaves, as well as fruit and seeds. They also store some of the food inside their trunks.

So ... all a tree needs to make food is water, air and sunshine.

Plants are living things, and all living things need energy to stay alive. Humans and other animals get their energy by eating food, but trees and other green plants make their food inside themselves. The way that plants make food is called "photosynthesis" (foe-toe-sin-theh-sis).

In the dry season, when the savannah had very little rain, the soil dried up. The tree and the other plants had no water. They struggled to make food and stay alive.

I'M ABSOLUTELY PARCHED!

The African savannah is hot all year round. It does not have hot summers and cold winters. Instead, it has dry seasons, when almost no rain falls, followed by wet or rainy seasons, when lots of rain falls.

The leaves on the tree turned brown and fell to the ground. The tree had to live on the food stored in its trunk.

When the rainy season arrived, the grasses and other plants grew tall and strong. The tree made new leaves that were green and glossy.

RAIN, GLORIOUS RAIN ...

Without plants, animals could not live in the savannah. Grasses, shrubs and trees make shady places where animals can shelter from the hot sun. But that isn't all — the plants also provide the animals with food.

The tree was full of insects busily feeding on it. Sugary sap from the green shoots was food for all kinds of bugs and ants, and crawling caterpillars munched on the leaves.

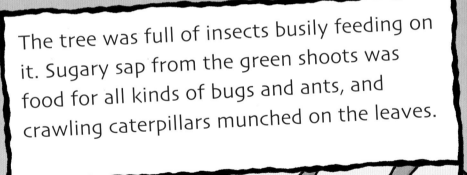

SLURP!

CHOMP!

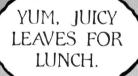

YUM, JUICY LEAVES FOR LUNCH.

All sorts of animals eat plants, from insects to elephants. Plant food gives the animals energy and the minerals their bodies need to be healthy. Plant-eating animals are called "herbivores".

A large caterpillar worked its way towards a clump of juicy leaves at the end of a twig. It did not notice the brightly coloured bird sitting on a branch above it.

YUM, JUICY CATERPILLAR FOR LUNCH!

Animals that eat other animals are known as "carnivores". As well as energy, carnivores also get minerals from the food they eat. Some animals (including humans) eat both plants and other animals — they are called "omnivores".

The bird bent its head to look at the caterpillar. In spite of its hairy spikes and orangey-red stripes, the bird knew it was good to eat.

In a flash, the bird reached down and grabbed the caterpillar in its beak. It flicked back its head and swallowed the caterpillar whole.

DOWN IN ONE!

Then the bird shook out its wings and settled back on the branch to wait for another meal to come along.

Every living thing, whether
it is a plant or an animal, is food
for something else.

Most animals only eat certain
types of food.

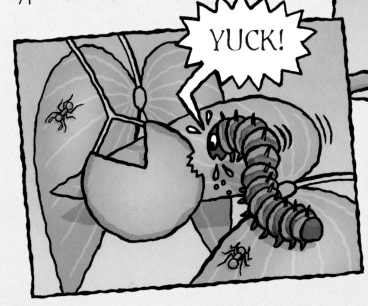

The caterpillar that eats the
leaves on the tree will not eat
fruit or seeds, even though they
grow on the same tree.

The bird that eats the
caterpillar might also eat other
insects, or even small lizards,
but it will not eat leaves.

Animals and plants are linked
together by the food that they
eat. Scientists call this a "food
chain". Each link of the chain is
food for the next link.

The first link in every food chain is always a plant. Plants are the only living things that make their own food. They are the food producers.

PRODUCER

A tree uses sunlight to make food.

CONSUMER

The tree leaves are eaten by a caterpillar.

CRUNCH!

CONSUMER

SLURP!

The caterpillar is eaten by a bird.

CONSUMER

The bird may be eaten by a bigger bird.

Animals are food consumers. They rely on the previous link in the chain to stay alive.

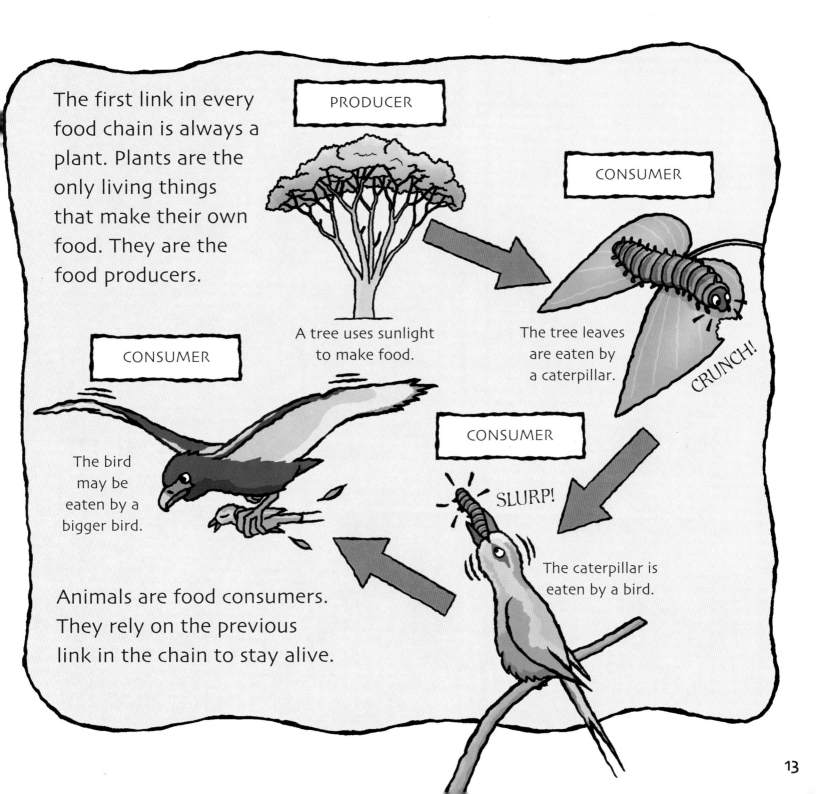

As the afternoon wore on it grew a little cooler and more animals came out to feed. The bird spread its wings and took to the air. A shadow fell on it from above. The shadow was made by an eagle.

EEEK!

With a loud shriek, the bird just managed to escape the eagle's outstretched claws. The bird darted away, its wings flapping wildly.

Disappointed, the eagle glided on. Below it, a small herd of zebra wandered across the grassland.

Animals that hunt and kill other animals are called "predators" — and the animals they hunt are known as "prey".

As the zebras moved, their striped coats made it hard to separate one animal from another. One zebra was limping and the rest of the herd travelled slowly so that it was not left behind.

ALL CLEAR!

Some people think a zebra's stripes help protect it from predators. When zebras bunch together in a group, their stripes overlap and the outline of each animal is harder to see. This makes it difficult for a predator to pick out any one animal.

The zebras stopped and lowered their heads to the grass. All except one. That zebra was on lookout. It kept its head up and swivelled its ears in all directions, checking for predators. But it did not see the lions hidden in the long grass.

The lionesses (female lions) had spotted the limping zebra. Keeping their bodies low to the ground they slowly spread out around the herd.

TIME FOR SUPPER.

The lionesses knew they had one chance to catch their prey. They had to get as close to the herd as possible before any of the zebras noticed them.

SNORT!

Suddenly another zebra raised its head and snorted nervously. In an instant, one of the big cats leapt forward, and the herd bolted in alarm.

Lions can run fast, but only for very short distances. If they don't catch their prey quickly then it will usually be able to outrun them.

For a moment the limping zebra was lost in the blur of zig-zagging stripes. But it could not run as fast as the others.

All the lionesses joined the chase. They drove the unlucky zebra away from the herd. One lioness jumped at the zebra's neck and another at its back — and they dragged it down. Within moments the zebra was dead.

WAIT FOR ME!

The two male lions in the group came forward, and the lionesses dropped back to let them feed first. Then the females ate, and finally the cubs were allowed to feed.

A group of lions is called a "pride". There are always many more females than males in a pride. The females usually do the hunting. The males protect the pride from other lions.

WHEN IS IT OUR TURN?

MUNCH!

SCRUNCH!

While the lions feasted, a flock of large brown birds began to gather. They were vultures — part of the savannah's clean-up crew.

The vultures waited impatiently for the lions to finish. When all the lions had eaten their fill, they padded away to find a place to rest. They would not need to hunt again until tomorrow.

At once the birds rushed in. Pushing and squabbling, they fought each other to get at the lions' leftovers.

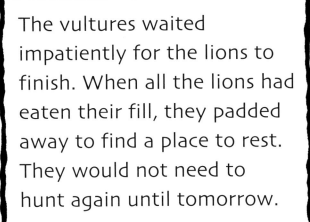

More vultures landed, but the noise and movement brought other visitors, too. A couple of hyenas came to see what the vultures had found, and they wanted a share.

Snarling and snapping, the hyenas chased the vultures away. Then they settled down to finish off what was left of the zebra. They crunched up the bones and even the hooves with their strong jaws. Soon there was little left but scraps of skin and bone.

Vultures and hyenas are both scavengers, although hyenas often hunt for their food as well. Scavengers eat what is left of another animal's kill. They also eat the bodies of animals that have died naturally of sickness or old age. In this way, no food is wasted.

CRUNCH!

CRACK!

The hyenas trotted off to search for something else to eat. The patch of grassland was quiet again — but it was not as empty as it seemed.

Thousands of insects scurried about in the grass. Some fed on the zebra scraps. Others laid eggs on them. When the eggs hatched, the scraps would provide food for the grubs.

As the days passed, the remaining bits of the zebra's body softened and rotted into the soil. Soon there was nothing to show that the zebra had ever been there. But even these rotted bits had one last purpose.

In the soil, invisibly small living things called bacteria were hard at work. Bacteria break up dead animal and plant material into minerals and other chemicals.

MMMM! THIS WATER IS GOOD.

The last bits of the zebra's body were finally used up. The next time it rained, water soaked into the soil and the minerals dissolved in the water.

Plant roots sucked up the mineral-rich water and made healthy new stems, leaves, seeds and fruit.

The plants were ready to start more food chains.

All living things produce waste material, such as dead leaves or dung. When plants and animals die, their bodies become waste material, too. All this waste material is broken down by other living things and used as food. As the waste is broken down, important chemicals are released into the air and soil. These chemicals are used by plants to help them grow. Then the plants are eaten by animals and the whole process starts again.

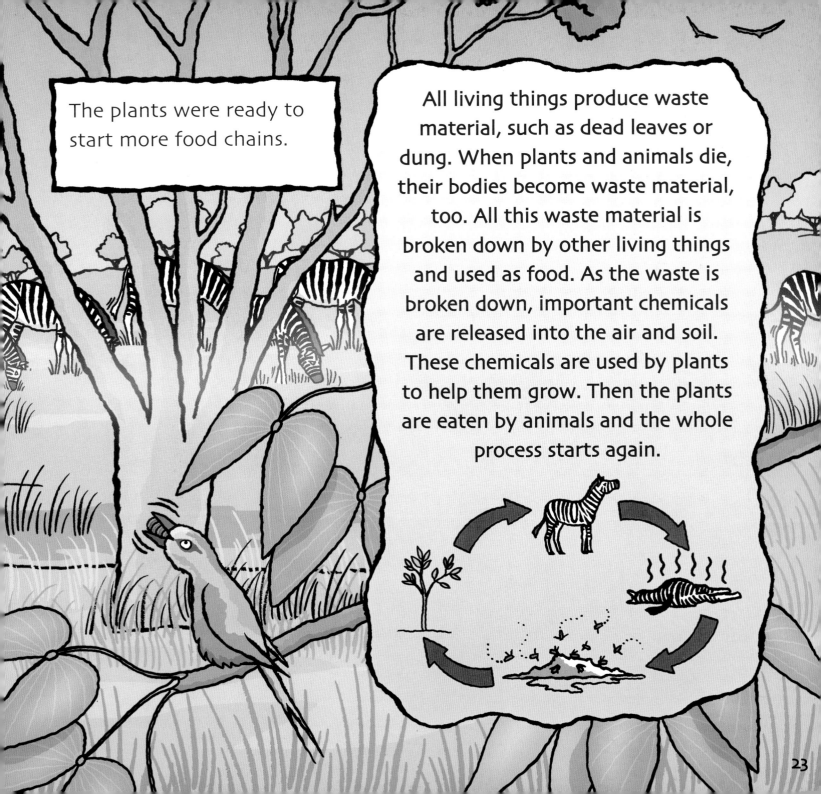

HOME, SWEET HOME!

THERE'S NO PLACE LIKE HOME

The place where an animal or a plant lives and grows best is called its habitat. The habitat gives the plant or animal the right amount of light or food, warmth or shelter, and a place where it can reproduce. A habitat can be big, like a savannah, or small, such as a pond.

Most living things can only survive in the right kind of habitat. A plant that grows in the hot savannah cannot live on a cold mountain. A caterpillar that lives in a tree cannot live in a pond.

CAUGHT IN A WEB

Many habitats are shared by lots of different plants and animals. Some of these plants and animals belong to more than one food chain.

The leaves that feed a caterpillar, for example, may also be eaten by a giraffe. The eagle that missed the bird will happily eat a lizard instead. Lions eat antelopes and giraffes, as well as zebras.

When food chains are linked together, they make a food web.

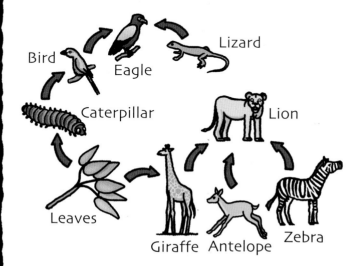

Bird

Eagle

Lizard

Caterpillar

Lion

Leaves

Giraffe   Antelope   Zebra

POISONED LINKS

If something happens to one part of a food chain it can affect all the other links in the chain.

For example, people sometimes spray plants with chemicals called pesticides. The pesticides kill the insects that feed on the plants. But other animals may eat the sprayed plants too, or the insects that feed on them. The poison in the pesticides is passed on from animal to animal. If it does not kill an animal straight away, it can build up in its body. If a predator eats enough poisoned prey, it will eventually die. Many eagles and other birds have died in this way.

TOP CAT

If plants are at one end of a food chain, a top predator is at the other end. Top predators are too powerful to be hunted by other animals as food. Instead, they die of illness or old age and their bodies are eaten by scavengers. Top predators are often fierce carnivores, such as lions and tigers. However, the most powerful top predators of all are not wild animals — they're humans!

I'M KING OF THE FOOD CHAIN ... I THINK!

25

TRY IT AND SEE

GROWING POWER

Most green plants start life as seeds. A seed contains everything it needs to make a new plant, but the new plant will only grow when it has the right conditions.

Try this experiment to find out what seeds and plants need to grow.

You will need:
- A clean bowl of water
- A teaspoon
- Seeds (mustard and cress, or sprouting seeds, such as alfalfa or lentils)
- Four clean, empty yogurt pots
- Cotton wool
- Water
- A shoebox
- Clingfilm

1 Put a teaspoon of seeds into the bowl of water and leave them to soak overnight. (You can miss this step if you are using mustard and cress.)

2 Number the yogurt pots 1, 2, 3 and 4. Put a thick layer of cotton wool into each yogurt pot. Sprinkle the cotton wool with water until it is damp.

3

Scatter a few seeds on top of the damp cotton wool in each pot. Put the pots somewhere warm and light.

4 Check the pots each day. Make sure the cotton wool does not dry out, but don't swamp it with water either. After a few days you should see little white shoots and roots coming from your seeds. Keep watering them until you see lots of green leaves appearing on the shoots. This should take about seven days.

5 Your seeds have grown into small plants. Now let's see what happens to them when you change their conditions for growth.

Leave Pot 1 where it is, and go on watering it each day.

Leave Pot 2 where it is, but don't give the plants any more water.

Put Pot 3 inside the shoebox and put the lid on the box. Water the plants as before, but keep the lid on the box so that they get no light.

Fill Pot 4 to the brim with water, and put clingfilm over the top so that the plants get no air.

How different are the plants in your pots after another seven days?

CONSUMING FACTS

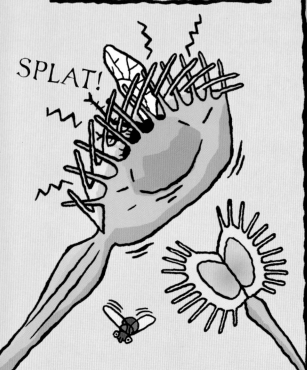

SPLAT!

There are a few plants that eat animals. They are called "carnivorous plants" and they mainly eat insects. One of the best-known is the Venus flytrap. When a fly lands on its hinged leaf, the leaf snaps shut, trapping the fly. The plant then dissolves the fly and sucks up the goodness in it.

Lots of animals like to eat the large caterpillars that live on the mopane tree in the African savannah — and people do, too. Called mopane worms, the caterpillars are a favourite snack food of the people who live in the savannah. They are eaten fried, dried, or stewed in a sauce.

OH NO, NOT ANOTHER ONE!

Hunting is difficult work, and lions often go hungry. A lion needs to eat about 8 kilograms of meat a day — that's as much as a large turkey. But if it gets the chance, it can eat six times as much as this.

INDEX

SOME FOOD CHAIN WEBSITES TO VISIT

http://www.bbc.co.uk/schools/revisewise/science/living/03b_act.shtml = an animated explanation of food chains and how they work.

http://ology.amnh.org/biodiversity/index.html = the American Museum of Natural History's website for kids, called Ology. The Biodiversity bit is full of great stuff on how living things behave together.

http://www.zephyrus.co.uk/landfoodchain.html = an interactive educational website containing information, puzzles and worksheets.

For Chris
JB
For David and Elizabeth
ML

First published in 2006 by
A & C Black Publishers Limited
38 Soho Square London W1D 3HB
www.acblack.com

Created for A & C Black Publishers Limited by

two's COMPANY

Copyright © Two's Company 2006

The rights of Jacqui Bailey and Matthew Lilly
to be identified as the author and the illustrator of this
work have been asserted by them in accordance with
the Copyrights, Designs and Patents Act 1988.

ISBN-10: 0-7136-7357-5 (hbk)
ISBN-13: 978-0-713-67357-9 (hbk)

ISBN-10: 0-7136-7358-3 (pbk)
ISBN-13: 978-0-713-67358-6 (pbk)

Printed and bound in China by Leo Paper Products

A & C Black uses paper produced with elemental chlorine-free
pulp, harvested from managed sustainable forests.